Fatma Sarsu • Abdelbagi M. A. Ghanim •
Priyanka Das • Rajeev N. Bahuguna •
Paul Mbogo Kusolwa • Muhammed Ashraf •
Sneh L. Singla-Pareek • Ashwani Pareek •
Brian P. Forster • Ivan Ingelbrecht

Pre-Field Screening Protocols for Heat-Tolerant Mutants in Rice

Fatma Sarsu
Plant Breeding and Genetics Section,
Joint FAO/IAEA Division
International Atomic Energy Agency
Vienna, Austria

Abdelbagi M. A. Ghanim
Plant Breeding and Genetics Laboratory,
Joint FAO/IAEA Division
International Atomic Energy Agency
Vienna, Austria

Priyanka Das
School of Life Sciences
Jawaharlal Nehru University
New Delhi, India

Rajeev N. Bahuguna
School of Life Sciences
Jawaharlal Nehru University
New Delhi, India

Paul Mbogo Kusolwa
Sokoine University of Agriculture
Morogoro, Tanzania

Muhammed Ashraf
Nuclear Institute for Agriculture and Biology (NIAB)
Faisalabad, Pakistan

Sneh L. Singla-Pareek
Plant Molecular Biology Group
International Centre for Genetic
Engineering and Biotechnology
New Delhi, India

Ashwani Pareek
School of Life Sciences
Jawaharlal Nehru University
New Delhi, India

Brian P. Forster
Plant Breeding and Genetics Laboratory,
Joint FAO/IAEA Division
International Atomic Energy Agency
Vienna, Austria

Ivan Ingelbrecht
Plant Breeding and Genetics Laboratory,
Joint FAO/IAEA Division
International Atomic Energy Agency
Vienna, Austria

Open Access provided with a grant from the International Atomic Energy Agency

https://doi.org/10.1007/978-3-319-77338-4

Preface

Global warming has potentially a huge negative impact on crop production. High temperatures have a direct damaging effect on crop development and yield, and locations where crops suffer from high temperatures have been identified worldwide. Rice is a major crop providing food for half of the world's population, and rice yield losses due to high temperatures have already been reported in many countries such as Australia, Bangladesh, China, India, Japan, Pakistan, the Philippines, Thailand, and the USA. Short-term predictions indicate that rice production could decrease by 10–25% in the near future because of higher temperatures.

Breeding heat-tolerant rice is one of the strategies to develop crop adaptation to the effect of climate change, particularly in major rice growing regions that are vulnerable to increased temperature. Developing high temperature-tolerant rice varieties is already an important breeding target for several national/international breeding programmes; however, changing weather patterns have increased the urgency to develop heat stress-tolerant rice varieties, particularly varieties that are already well adapted to local environments.

Mutation breeding is an effective approach to develop heat stress tolerance in crops, including rice. Therefore, rice mutation breeding for adaptation to high temperatures can augment current technology to maintain crop yields. The traditional approach of screening for heat stress tolerance under field conditions is hampered by the unpredictability of field and weather conditions and therefore presents challenges to plant breeders who typically must screen large populations to detect rare, useful variants.

To meet the increasing demands from countries for heat stress-tolerant crops and to help address the effects of climate change on agricultural production, the Plant Breeding and Genetic Section of the Joint FAO/IAEA Division of Nuclear Techniques in Food and Agriculture launched the Coordinated Research Project (CRP) 23029 on 'Climate Proofing of Food Crops: Genetic Improvement for Adaptation to High Temperatures in Drought Prone Areas and Beyond', which ran from 2011 to 2016. This book is the result of this CRP, and its main purpose is to provide robust, user-friendly protocols for effective pre-field screening of mutant rice for enhanced heat stress tolerance. Because mutation breeding involves the screening of large mutant populations, effective protocols are required to reduce the cost and labour of selecting the rare, useful variants. The protocols are described for the seedling and the flowering stages, the two critical development stages most vulnerable to increased temperatures.

The CRP indicates that it is possible to simplify the identification of heat-tolerant lines of rice among breeding populations in glasshouse and controlled-environment growth chambers using a screening method at the seedling stage. These protocols are primarily designed for use by plant breeders who need practical and rapid screens to process large mutant populations, including segregating populations, advanced generations and rice germplasm collections. They may also be more widely adapted to screen for heat stress tolerance in other crops.

The protocols were developed at the FAO/IAEA Plant Breeding and Genetics Laboratory in Seibersdorf, Austria, in collaboration with experts in Cuba, India, Pakistan, and the United Republic of Tanzania, who were also been involved in the field testing and physiological studies required to validate the results of the pre-field screening protocols.

Vienna, Austria Fatma Sarsu

The original version of this book was revised. The correction is available at https://doi.org/10.1007/978-3-319-77338-4_5

Acknowledgements

We would like to thank all participants of CRP 23029 for testing and validating the protocols in their rice mutation breeding programmes as well as for their valuable insights and feedback. We also thank the National Institute of Agricultural Sciences (INCA), Cuba, for sharing with us field testing results of mutant lines and cultivars. We thank the Soil and Water Management and Crop Nutrition Laboratory of the Joint FAO/IAEA Division for providing the necessary infrastructure.

We also thank the following external reviewers for their invaluable input to this book: Chen Zhiwei, College of Crop Sciences, Fujian Agriculture and Forestry University, China; Maria Caridad Gonzales Cepero, INCA, Cuba; Kanchana Klakhaeng, Rice Department, Thailand; Necmi Beser, Trakya University, Turkey; and Sudhir K Sopory, International Centre for Genetic Engineering and Biotechnology, India.

Executive Summary

In this publication, we present simple, robust pre-field screening protocols that allow plant breeders to screen for enhanced tolerance to heat stress in rice in a breeding programme using a controlled environment. Two critical heat-sensitive stages in the life cycle of the rice crop were targeted: seedling and flowering stages with screening based on simple phenotypic responses. The protocols are based on the use of a hydroponics system and/or pot experiments in a glasshouse in combination with a controlled growth chamber where the heat stress treatment is applied. The protocols are designed to be effective, simple, reproducible, and user-friendly.

The methods include a protocol for screening heat tolerance of rice at the seedling stage; young seedlings were exposed to heat stress of 45 °C/28 °C for 6/18 h during 4–6 days with 80% relative humidity. The seedling test takes 4–5 weeks and involves the visual scoring of symptoms which allows hundreds of seedlings to be evaluated in a short time. The visual screening method was extensively validated through laboratory, glasshouse, and field-based experiments. Heat stress tolerance was assessed according to a heat tolerance index, which is based on seedling biomass using shoot, root, and whole seedling weight (fresh and dry) parameters as well as root and shoot length and seedling height. The seedling test can be used to screen M2 populations, advanced mutant lines as well as cultivars. We also adapted a protocol for screening heat-tolerant mutant lines at the flowering (reproductive) stage that has been specifically adjusted for a mutation breeding programme. Here, plants were treated from the first day of anthesis at different temperatures at 35.0–39.0 °C/28 °C for different durations 6–4/18–20 h for 6–4 days and 80% relative humidity. Spikelet fertility at maturity was determined as a parameter to assess the heat tolerance of the selected genotypes.

Selected heat-tolerant mutant rice genotypes were tested for physiological and biochemical indicators associated with the pre-field screen protocols. These tests included measuring physiological and biochemical indicators associated with plant stress responses, such as electrolyte leakage, malondialdehyde level, total protein content, and antioxidant enzyme activity at seedling, vegetative, and flowering stages to understand the mechanism of the heat tolerance characteristics/traits of the selected germplasm and explore the potential of pyramiding different mutations for durable heat tolerance.

Furthermore, the candidate heat-tolerant mutant lines were also tested in hot spot areas in the field in Cuba, Pakistan, and the United Republic of Tanzania to evaluate their performance under field conditions in heat-stressed growing environments. About 70% of the mutants that were pre-selected at the seedling stage in an environmentally controlled growth chamber also showed heat stress tolerance under field conditions, thus validating our pre-field screening protocols.

Overall, the field and laboratory (physiological and biochemical) testing of these genotypes shows that the developed pre-field screening protocols can reliably identify mutant rice lines with enhanced heat stress tolerance also under field conditions.

The protocols described here will enable plant breeders to effectively reduce the number of plants from a few thousands to less than 100 candidate individual mutants or lines in a greenhouse/growth chamber for further testing in the field conditions using replicated trials. In addition, the methods can also be used to classify rice genotypes according to their heat tolerance characteristics. Thus, different types of heat stress tolerance mechanisms could be identified offering opportunities for pyramiding different (mutant) sources of heat stress tolerance.

Contents

Chapter 1
General Introduction

1.1 Background Analysis

Global warming has become a serious problem affecting agricultural production in vulnerable regions worldwide and is projected to worsen with anticipated climate change. Rice (*Oryza sativa L.*) is a major (staple) food crop associated with the lives of three billion people around the world. It is planted on about 159 million hectares annually in at least 114 countries by more than 100 million households in Asia and Africa (Tonini and Cabrera 2011). Rice is the source of 27% of dietary energy and 20% of dietary protein in the developing world and rice is the major staple crop for nearly half of the world's population (Mottaleb et al. 2012).

Global environmental projections forecast that during the twenty-first century, global surface temperatures are likely to rise by 1.1 to 2.9 °C for the lowest carbon emission scenarios, and by 2.4 to 6.4 °C for the highest emission scenarios (IPCC 2012). The increase in temperature can cause irreversible damage to plant growth and performance, with major consequences on crop yield and also quality (Wahid et al. 2007). A 7–8% reduction in rice yield is associated with each 1 °C rise in day temperature from 28 to 34 °C (Baker et al. 1992). Using yield data from field experiments, Peng et al. reported that rice yields decline with higher night temperature from global warming. Increased temperatures cause reductions in the rate of photosynthesis and stomatal conductance at all growth stages in the life cycle of rice, at both vegetative and reproductive stages (Sanchez-Reinoso et al. 2014; Yoshida 1981).

Rice has been cultivated under a broad range of climatic conditions. Around 90% of the global rice crop is grown and consumed in Asia, where 50% of the population depends on rice as a regular and daily food (Pareek et al. 2010). In Asia the rice crop is particularly vulnerable to high temperatures (above 33 °C) during the sensitive flowering and early grain-filling stages (Wassmann et al. 2009a, b).

Regional high temperature damage to rice crops was also documented in many tropical and sub-tropical countries, such as Pakistan, India, Bangladesh, China,

F. Sarsu et al., *Pre-Field Screening Protocols for Heat-Tolerant Mutants in Rice*,
https://doi.org/10.1007/978-3-319-77338-4_1

Thailand, Sudan and some other African countries (Osada et al. 1973; Matsushima et al. 1982; Li et al. 2004; Xia and Qi 2004; Yang et al. 2004; Tian et al. 2009). The problem is the most acute when temperature extremes coincide with critical sensitive stages in crop development. Heat tolerance for a crop is generally defined as the ability of plants to grow and produce an economic yield under high temperature (Wahid et al. 2007). Heat stress creates a serious threat to rice production, including in the most productive regions of the world, and it is imperative that heat tolerance is included as a target trait in breeding new rice cultivars (Pareek et al. 2010). In general, the reproductive stage is more vulnerable to heat stress than the vegetative stage in many crop species. In rice, almost all growth stages are affected by high temperature. During the vegetative growth period, rice can tolerate relatively high temperatures up to 35 °C. Temperatures above this level may reduce plant growth, flower initiation and ultimately yield. High temperatures are particularly damaging if they occur at the time of anthesis, and pollen shedding (Yoshida 1981). The two most sensitive stages are seedling stage, booting (microsporogenesis) and flowering (anthesis and fertilization). High temperature affects plant growth, meiosis, anther dehiscence, pollination, and pollen germination, which leads to spikelet sterility and yield loss (Yoshida 1981; Wassmann et al. 2009a, b; Shah et al. 2011; Prasanth et al. 2012; Tenorio et al. 2013; Sanchez-Reinoso et al. 2014). Exposure of rice plants to temperatures above 35 °C for short periods, less than one hour, during anthesis may result in varying degree of pollen and spikelet sterility which leads to significant yield losses and low grain quality (Jagadish et al. 2007; Matsui et al. 1997; Ye et al. 2015). Thus spikelet fertility under high temperature has been widely used as a screening index for heat tolerance at the flowering stage (Ye et al. 2015). Also the time of anthesis can affect sterility with early morning anthesis preferred since high temperature is avoided and hence high temperature induced sterility is reduced (Yoshida 1981).

Yield stability can only be improved if a breeding program is based on the valuable new knowledge on plant development and stress responses (Barnabas et al. 2008). A critical step in plant breeding is the ability to screen for the trait of interest. This study set out to develop simple, but effective protocols to screen for rare heat tolerant rice mutant plants among a host of non-improved siblings.

1.2 Physiology and Genetics of Heat Tolerance in Rice

Plant heat stress tolerance can be sub-divided into (1) escape; successful reproduction before the stress, such as the timing of panicle emergence and spikelet/floret opening before the occurrence of the stress (Singla et al. 1997), (2) avoidance; maintenance of a cooler canopy with higher transpiration from leaf surface, (3) true tolerance which may involve various physiological mechanisms induced under the stress (Kondamudi et al. 2012; Bahuguna et al. 2015; Bahuguna and Jagadish 2015). In rice a short exposure of seedlings to high temperature can affect the plant cellular ultrastructure with major changes occurring in the chloroplasts and mitochondria, thus resulting in reduced metabolism and hence also reduced growth (Pareek et al. 1997).

With respect to physiology, plants can adjust their metabolism and morphology in response to heat stress (Singla et al. 1997). High temperatures generally induce the expression of heat shock proteins (HSPs) and suppress, at least in part, the synthesis of normal cellular protein production (Shah et al. 2011). Heat shocks proteins (HSPs) are induced in response to short-term stress but may also be important to adapt to the heat stress (Pareek et al. 1995). HSPs can improve or stabilise photosynthesis, partitioning of assimilates, nutrient and water use efficiency and the thermal stability of cellular membranes (Wahid et al. 2007). Some of these HSPs and molecular chaperones aid in restoring damaged proteins (Kumari et al. 2013). These mechanisms need to be investigated further in agricultural production systems if they are to be exploited in developing heat stress-tolerant rice cultivars (Sailaja et al. 2015).

The genetics of heat tolerance is poorly understood, but is complex and controlled by multiple genes (Wahid et al. 2007; Xue et al. 2012; Driedonks et al. 2016). Heat tolerance in rice has a fairly high heritability and most genetic variation is additive (Yoshida 1981). There is huge variation for heat stress tolerance in rice as cultivars, lines and genotypes have been reported which are sensitive, tolerant or intermediate in response (see Ye et al. 2015). Many HSPs have been reported and their genetics (controlling genes, location of genes, dominance/recessive-ness) are known. However, certain gene combinations are critical to successful cultivar breeding, e.g. cultivars have to have the optimal genes/alleles for flowering time, height, etc., and it is not known how effective HSP genes are in an elite genetic background.

According to recent genetic studies plant heat-tolerance is probably a polygenic trait. In wheat different components of tolerance, controlled by different sets of genes, are critical for heat tolerance at different stages of development or in different tissues (Barakat et al. 2011). Shah et al. (2011) emphasized that *indica* rice is generally more heat tolerant than *japonica* rice, however there is a genotypic variation in spikelet fertility at high temperature in both species. Understanding the genetic basis of tolerance and enhancing the breeding level of heat tolerant cultivars in rice still continue. In rice, the development of molecular marker technology has led to the identification of several QTL for heat tolerance (Xue et al. 2012). It is known that the mapping populations and accurate phenotyping technology are essential for QTL mappings (Zhao et al. 2016). According to Zhong-Hua et al. (2014), after discovery of mutation through phenotyping, the mutant can be used for gene discovery. Thus far, 64 genes which are responsible for mutant phenotypes photosynthesis, signalling transduction and disease resistance have been isolated and mapped to the rice genome. Heat tolerance in rice at the flowering stage is controlled by several QTLs. Pyramiding validated QTL's for heat tolerance QTL's could be an important mechanism to enhance heat tolerance in rice at flowering stage, focused in spikelet fertility (Ye et al. 2015). They confirmed that the presence of recessive QTLs on chromosome 4 results in 15% higher fertile rice spikelet compared to plants without the QTL. Moreover Zhao et al. (2016) stated that using marker assisted selection (MAS) breeding strategy is essential, although many of putative QTLs for heat tolerance at anthesis have been identified, the effect and stability of the target QTL needs to be further confirmed.

The completion of the Rice Genome Sequencing Project and high-throughput genotyping and phenotyping have generated valuable data and tools that can be used to identify genes associated with target traits such as heat tolerance (Zhong-Hua et al. 2014). These advances will facilitate the dissection of genetic controls of heat tolerance in rice that may then be exploited in the development of new heat tolerant rice varieties.

1.3 Physiological and Biochemical Heat Stress Indicators in Rice

Heat stress alters a wide range of physiological, biochemical and molecular processes affecting crop growth and yield (Mittler et al. 2012; Hasanuzzaman et al. 2013). Photosynthesis is highly sensitive to heat stress, and above 35 °C decreases by 50% in rice. Another major physiological consequence of heat stress is augmented levels of reactive oxygen species in cells, which leads to oxidative stress (Hasanuzzaman et al. 2013). Plants can tolerate sub-lethal heat stress by avoidance, escape or physical changes at the cellular level such as changing membrane physical state, the synthesis of specialized HSPs and augmented anti-oxidative defence (Mittler et al. 2012; Bahuguna and Jagadish 2015). Plants acclimate to sub lethal heat stress by altering metabolism at physiological, biochemical and molecular levels. Changes in the membrane physical state and composition, production of heat shock proteins, transcription factors, osmolytes and augmented levels of anti-oxidant defence are key processes to maintain cellular redox homeostasis under heat stress (Krasensky and Jonak 2012). Heat stress alters gene expression patterns (Shinozaki and Yamaguchi-Shinozaki 2007) leading to the acclimation and/or adaptation to heat stress with improved antioxidant defence and higher levels of heat shock proteins, which can protect the integrity of proteins and other biological molecules (Moreno and Orellana 2011). Moreover, plants can modify their metabolism in various ways in response to heat stress, notably by generating compatible solutes that are able to organize proteins and cellular structures, maintain cell turgor pressure, and modify anti-oxidant mechanisms to re-establish cellular redox homeostasis (Munns and Tester 2008; Janska et al. 2010). Heat stress also alters gene expression which involves 'direct protection' from high temperature stress (Shinozaki and Yamaguchi-Shinozaki 2007). These proteins include osmoprotectants, transporters, anti-oxidants and regulatory proteins (Krasensky and Jonak 2012). Moreno and Orellana (2011) indicated that in heat stress, alterations in physiological and biochemical processes caused by gene expression progressively lead to the development of heat tolerance in the form of acclimation and/or adaptation, but this may not be associated with yield.

Key physiological and biochemical indicators of heat stress tolerance include electrolyte leakage, lipid peroxidation level i.e. malondialdehyde (MDA) content and anti-oxidant enzyme activity (Campos et al. 2003; Heath and Packer 1986; Bajji et al. 2001; Nakano and Asada 1981; Oberley and Spitz 1985). High temperatures may also affect membrane stability through lipid peroxidation, leading to the production of peroxide ions and MDA. A change in concentration of MDA is a good indicator of membrane structural integrity under temperature stress (Sanchez-Reinoso et al. 2014). Increase temperature stress 37 °C/30 °C (day/night) increased MDA content and electro leakage percentage in rice (Zhang et al. 2009; Liu et al. 2013).

1.4 Breeding for Heat Tolerance in Rice

Induced mutation has been hugely successful in rice breeding and could augment ongoing breeding efforts for enhancing heat stress tolerance in rice and other crops (Forster et al. 2014). Thus far, 824 rice cultivars have been developed by mutation breeding using mostly gamma irradiation, but also Ethyl methane sulfonate (EMS) and fast neutron (IAEA 2017 mutant variety database). One significant example is the first semi-semi-dwarf dwarf rice cultivar, Calrose 76, developed with 15% higher yield than taller cultivars, and it has also been used as a source of many semi-dwarf dwarf cultivars by rice scientists (Zhong-Hua et al. 2014). Another significant example is Zhefu 802 mutant rice variety was grown in China 10.6 million ha from 1986 to 1994 in China (Shu et al. 1997).

Although heat-tolerant rice genotypes have been found in both sub-species (Prasad et al. 2006) it was noted that *indica* spp. are more tolerant to higher temperatures than *japonica* spp. (Satake and Yoshida 1978). Recently, new heat stress tolerant rice cultivars have been generated by conventional cross breeding; examples include heat-tolerant lines and released cultivars such as NH 219, Dular, Nipponpare and WAB56-125. These are popular heat tolerant cultivars in South East Asia, particularly in the Philippines, Vietnam, Thailand, Indonesia and Cambodia (Poli et al. 2013; Manigbas et al. 2014). Moreover high quality Pon-Lai rice developed through cross breeding with a parent *japonica* type good quality mutant variety and another parent *indica* type heat stress tolerant variety to breed heat tolerant and high quality rice in Taiwan (Wu et al. 2016). In addition five heat stress tolerant *japonica* cultivars were bred from 2005 to 2011 in Japan (Takahashi et al. 2016).

To date, the *indica* rice genotype N22, an EMS induced mutant which is a deep-rooted, is the most tolerant genotype for heat stress and drought (Yoshida et al. 1981; Prasad et al. 2006; Poli et al. 2013). Many studies have demonstrated genotypic variation in spikelet sterility at high temperatures (Satake and Yoshida 1978; Prasad et al. 2006) and the fertility of spikelets at high temperature can be used as a screening tool for reproductive stage (Shah et al. 2011). A drawback of conventional breeding is that the programmes are often based on local elite lines with low genetic diversity (Driedonks et al. 2016) and consequently unlikely to possess variation for new traits such as heat tolerance. Wide crossing with more exotic material may provide the required genetic variation, but the breeding process will take longer to clean up the genetic background after the initial cross.

Induced mutation is a heritable change in the genetic material of living organisms, and this has been a major driver in species diversity and evolution. Plant breeding requires genetic variation of useful traits for crop improvement. The use of various mutagens to generate genetic variation in crop plants has a history almost as long as that of conventional breeding. Mutation breeding involves the development of new cultivars by generating new genetic variability induced by chemical and physical mutagens. Once mutation is produced the next steps are to detect and identify mutants with desired traits, i.e. screening. Mutant selection is a key step in a mutation breeding program, requiring the screening of thousands of mutants to recover the rare mutant with the desired trait. Hence, a major bottleneck in plant mutation breeding is effective screening of rare desired mutants having genetically improved characteristics among a mutant population comprising thousands of plants.

A major challenge in breeding for heat tolerance is the identification of reliable screening methods and effective selection criteria to facilitate detection of heat-tolerant plants. Several screening methods and selection criteria have been developed by different researchers (Wahid et al. 2007; Ye et al. 2015), but for practical plant breeding these need to be rapid and efficient in terms of time, space and cost. Due to the complexity of heat stress, there is need to develop quick and fast screening protocols for heat tolerance and plant breeders are still in need for identifying such efficient screening tools for detecting heat tolerance potentials at early growth stages in crops. Therefore, there is an urgent need for reliable pre-field screening protocols to enhance the efficiency and effectiveness of plant mutation breeding.

Heat stress alters a wide range of physiological, biochemical and molecular processes affecting crop growth and yield (Mittler et al. 2012; Hasanuzzaman et al. 2013). Photosynthesis is highly sensitive to heat stress, and above 35 °C decreases by 50% in rice. Another major physiological consequence of heat stress is augmented levels of reactive oxygen species in cells, which leads to oxidative stress (Hasanuzzaman et al. 2013). Plants can tolerate sub-lethal heat stress by avoidance, escape or physical changes at the cellular level such as changing membrane physical state, the synthesis of specialized HSPs and augmented anti-oxidative defence (Mittler et al. 2012; Bahuguna and Jagadish 2015). Plants acclimate to sub lethal heat stress by altering metabolism at physiological, biochemical and molecular levels. Changes in the membrane physical state and composition, production of heat shock proteins, transcription factors, osmolytes and augmented levels of antioxidant defence are key processes to maintain cellular redox homeostasis under heat stress (Krasensky and Jonak 2012). Heat stress alters gene expression patterns (Shinozaki and Yamaguchi-Shinozaki 2007) leading to the acclimation and/or adaptation to heat stress with improved antioxidant defence and higher levels of heat shock proteins, which can protect the integrity of proteins and other biological molecules (Moreno and Orellana 2011). Moreover, plants can modify their metabolism in various ways in response to heat stress, notably by generating compatible solutes that are able to organize proteins and cellular structures, maintain cell turgor pressure, and modify anti-oxidant mechanisms to re-establish cellular redox homeostasis (Munns and Tester 2008; Janska et al. 2010). Heat stress also alters gene expression which involves 'direct protection' from high temperature stress (Shinozaki and Yamaguchi-Shinozaki 2007). These proteins include osmo-protectants, transporters, anti-oxidants and regulatory proteins (Krasensky and Jonak 2012). Moreno and Orellana (2011) indicated that in heat stress, alterations in physiological and biochemical processes caused by gene expression progressively lead to the development of heat tolerance in the form of acclimation and/or adaptation, but this may not be associated with yield.

Key physiological and biochemical indicators of heat stress tolerance include electrolyte leakage, lipid peroxidation level i.e. Malondialdehyde (MDA) content and anti-oxidant enzyme activity (Campos et al. 2003; Heath and Packer 1986; Bajji et al. 2001; Nakano and Asada 1981; Oberley and Spitz 1985). High temperatures may also affect membrane stability through lipid peroxidation, leading to the production of peroxide ions and MDA. A change in concentration of MDA is a good indicator of membrane structural integrity under temperature stress (Sanchez-Reinoso et al. 2014). Increase temperature stress 37 °C/30 °C (day/night) increased MDA content and electro leakage percentage in rice (Zhang et al. 2009; Liu et al. 2013).

Chapter 2
Screening Protocols for Heat Tolerance in Rice at the Seedling and Reproductive Stages

2.1 Background Analysis

The optimum temperature for rice germination is between 28 and 30 °C. High temperature affects almost all growth stages of rice from germination to ripening (Shah et al. 2011). The threshold temperature at the seedling stage has been identified as 35 °C; the main symptom of heat stress is poor growth (Yoshida 1981).

Prasanth et al. (2012) tested heat stress in different stages of plant and used 28 genotypes including three mutant lines. These authors noted genotype-specific response with regards to heat stress tolerance in the three different stage analysed: germination, seedling and early vegetative stage. They added that there are fewer reports on the effect of high temperature on germination and vegetative stage of rice seedlings compared to the reproductive stage.

Plant reproductive processes are complex and sensitive to environmental changes, including high temperatures, which ultimately affect fertilization and post-fertilization processes leading to decreased yields. Flowering is one of the most susceptible stages in the life cycle of rice, and rice spikelets at anthesis exposed to more than 35 °C for 4–5 days induces sterility, with no seed produced (Satake and Yoshida 1978). Temperatures above 35 °C at flowering causes failure of anther dehiscence, and thus less pollen, resulting in incomplete fertilization in rice (Jagadish et al. 2007; Prasad et al. 2006; Satake and Yoshida 1978). Weerakoon et al. (2008) using a combination of high temperatures (32–36 °C) with low (60%) and high (85%) relative humidity recorded high spikelet sterility. Flowering (meiosis, anthesis and fertilization) is considered to be one of the most sensitive stages for temperature stress in rice. The threshold temperature for success in flowering in rice is 33 °C (Jagadish et al. 2007). Therefore, the fertility of spikelets at high temperature can be used as a screening tool for high temperature tolerance (Satake and

The original version of this chapter was revised. A correction to this chapter is available at https://doi.org/10.1007/978-3-319-77338-4_5.

F. Sarsu et al., *Pre-Field Screening Protocols for Heat-Tolerant Mutants in Rice*,
https://doi.org/10.1007/978-3-319-77338-4_2

Yoshida 1978). Some cultivars that are not so sensitive to higher temperatures, e.g. cv. N22, can be used in breeding programmes as controls in screening tests. Besides spikelet fertility occurring during the flowering stage, a phenotypic marker for male gametogenesis has been reported as a screening tool for high temperature stress in rice (Jagadish et al. 2014). Authors identified a distance of 8–9 cm between the collar of the last fully opened leaf and the flag leaf collar as the environmentally stable phenotypic marker to assess heat stress (38 °C) sensitivity at the micro sporogenesis stage leading to high spikelet sterility in rice.

Wei-hun et al. (2012) applied heat stress at 15 day-old seedling stage 43 °C/30 °C (day/night and 14 h/10 h), 75% relative humidity for 7 days and observed that plant height, root length, shoot and root biomass were dramatically reduced, with heat tolerant genotypes being relatively less affected than sensitive ones. In this sense, we have included seedlings as a target for heat tolerance screening, to enable selection at the seedling stage by plant breeders. This method, as described in Sect. 2.2, is based on growing rice seedlings in glasshouse conditions using hydroponics or pots and exposing them to heat stress in a growth chamber. The seedling test is simple and rapid (4–6 days) and allows the simultaneous screening of several hundred seedlings at once depending on the size of available infrastructure. A list of equipment required for hydroponics hardware and stock solutions is given. Heat stress treatment was applied after 15 days of seedling transplantation to hydroponics or pots. Germplasm classifications for sensitive, intermediate and tolerant are also provided. Visual symptoms of heat stress include: reduced leaf area, yellowing of leaves, leaf tip burning, and entire leaf burning and leaf death. Assessment of seedlings was done according to a heat tolerance index, which is calculated for seedling biomass data using shoot, root, whole seedlings weight (fresh and dry), root, shoot length and seedling height. The test can be adapted to screen M_2 individuals as well as M_3 families from mutant populations (and other large breeding populations). A stress tolerance index (STI, Fernandez 1992) was used to identify/validate genotypes producing high yield under both stress and non-stress conditions; Blum (1988) stated that selection based on STI can result in genotypes with higher stress tolerance and yield potential to be selected.

Tenorio et al. (2013) and Rang et al. (2011) developed a phenotyping protocol for use in plant breeding in which a temperature treatment of 38 °C is used at the flowering stage. Our second protocol presented here is to screen for heat tolerance in rice at the flowering stage modified from Tenorio et al. (2013) and Rang et al. (2011). Flowering is most heat sensitive stage that can result in high sterility and yield loss in rice. As explained in Sect. 2.3, glasshouse grown plants are moved to a controlled environmental from the first day of anthesis for heat stress treatments, at 35.0–39.0 °C/28 °C for 6–4/18–20 h with relative humidity 80% for 6–4 days. After heat treatment plants may be rescued and returned to glasshouse under optimum temperature conditions to produce seed and for further evaluation. Spikelet fertility (seed set) is measured as an important predictive parameter for yield (Prasad et al. 2006; Rang et al. 2011). In our protocol also we used spikelet fertility at maturity to assess the heat tolerance of the tested genotypes.

2.2 Screening Protocol for Heat Tolerance in Rice at the Seedling Stage

2.2.1 Plant Materials

Experiments were performed using controlled environment facilitates, test plant materials were compared to standard genotypes of known heat stress tolerance. Rice controls used at the PBGL (Austria), India, UR Tanzania and Pakistan are:

N22: a heat tolerant mutant, India origin

IR64: an intermediate heat tolerant genotype, from IRRI, Philippines

IR52: a heat susceptible genotype, from IRRI, Philippines

WAB 56-104: a heat susceptible genotype, Nigeria origin

The heat tolerance of the above standards in hydroponics and pots has been correlated with field performance, see Sect. 3.3. These standard materials can be requested free of charge from IRRI under a Standard Material Transfer Agreement. It is advisable to include local cultivars or breeding lines of known heat stress tolerance in tests.

2.2.2 Screening for Heat Tolerance Using Hydroponics

The equipment for hydroponics, the preparation of nutrient stock solutions and setting up of the hydroponics system are given in Bado et al. (2016).

2.2.2.1 Seedling Establishment in Hydroponics

The seedling establishment was conducted in a glasshouse set up for rice: 28/20 °C and 70% humidity in 16 h photoperiod. Test tanks were filled with distilled water until the water level is about 1 mm above the mesh of the platform. Dry seed were then placed into the wet compartments of the seed support platform. For M_2 screening 20–40 seeds from one panicle is placed into one compartment (6 × 7 cm) (Fig. 2.1a). For M_3 and advanced lines/cultivars 3–5 seeds were placed into each 2 cm diameter compartment. Individual lines may be replicated within and among tanks (Fig. 2.1b). The test platforms were then covered with a lid for 5–6 days to promote germination in the dark. On day 3, the seed should show signs of germination (emergence of root and shoot) and the water should be replaced with half-strength Yoshida solution as vigorous seedlings require nutrients. Germinated seeds were allowed to grow for 6 days. The seedlings for subsequent heat stress treatment should be healthy, therefore odd-looking or diseased seedlings should be removed at this stage. If seed samples are not clean and rotting occurs during germination, these must also be removed. If symptoms are widespread, a new batch of seedlings and fresh hydroponics materials should be prepared.

Fig. 2.1 Germination and seedling establishment for heat stress treatment in hydroponics. (**a**) Seed support platform with rice seeds ready for germination, each compartment contains seed from one panicle per M_2 plant. (**b**) M_3/M_4 advanced line seed support platform showing rice seeds ready for germination, each compartment contains 3–5 seeds from each line. After germination reduced to one seedling for each compartment (**c**) Rice seedlings are ready for heat application. (**d**) After heat treatment, rice seedlings showing various degrees of heat injury, some rows all died (last two rows), some could show tolerance with less damage. (Pictures from PBGL, Joint FAO/IAEA Division)

2.2.2.2 Care Required During Plant Growth in Hydroponics

Due to evaporation from tanks and transpiration from plants there will be a gradual decrease in solution volumes over time. Every 2 days the volume needs to be brought back to the level of full capacity (touching the netting in the platform compartments) and the pH adjusted to 5. If algal contamination is found, the nutrient solution should be completely replaced. Solutions can be changed by lifting off the platforms and placing them temporarily over empty tanks and pouring the hydroponics solutions back into a drum where the bulked solution can be pH adjusted for the whole experiment in one step. Once adjusted the solution is re-distributed into the test tanks and the seedling platforms returned. These operations also act to aerate the hydroponics solution. Alternatively, the pH can be adjusted on an individual tank basis and more working solution may be added to make up the volume in each tank.

2.2.3 Screening for Heat Tolerance in Soil Using Pots

No special equipment is needed for screening with soil in pots and/or trays. Plastic pots, trays or tanks are filled with a clay soil which has a good water retention capacity and organic matter content. Pre-germinated, 6 day-old young healthy seedlings are transferred to the plastic pots, trays or tanks as given in Fig. 2.2a–c. Also as shown in Fig. 2.4d, in case germination problems, an excess (110–120) of seeds of each line are sown directly into the tray to have 100 seedlings in a tray 1 week after sowing.

Germinated 6 day-old seedlings were transferred to pots or trays to grown in a glasshouse set up for rice: 28/20 °C day/night temperatures with 70% humidity and 16 h period. The glasshouse should be disease free and well-lit by natural or artificial lighting. Standard rice management practices are applied to pots with respect to soil, fertilizer and watering until the application of heat stress.

Fig. 2.2 Germination and seedling establishment for heat stress in trays. (**a**) Pre-germinated rice seedlings in petri dishes with filter paper. (**b**) Young seedlings transplanted into trays. (**c**) 15 days after transplantation seedlings are ready for heat treatment (**d**) 45 °C heat application in growth chamber (Pictures are from Sokoine University, URT)

2.2.4 Heat Treatment and Recovery of Seedlings

Fifteen days after transplantation, seedlings were transferred to the growth chamber to grow under tightly controlled temperature, light and relative humidity conditions. Growth chambers were adjusted to 45 °C/28 °C (day/night) 6/18 h with 80% relative humidity. The temperature treatment was imposed for 4–6 days; depending on seedling tolerance to heat stress application (some of them already died after 4 days). For all genotypes including control cultivars, 50/100 seedlings were exposed to heat stress and same amount of seedlings are grown in glasshouse conditions without stress as controls (Figs. 2.3 and 2.4).

Fig. 2.3 Heat treatment to seedlings in trays/hydroponics, each tray contains 100 seedlings in control and stress conditions (**a** and **c**) Control seedlings no heat stress applied, (**b**) After heat stress treatment, heat stress effects on seedlings most of the genotypes already dead, one to two lines alive seems tolerant to heat stress (**d**) Seedlings in hydroponics ready for heat stress application, each tank contains 100 seedlings. Right side: hydroponically grown seedlings showing various degrees of heat injury, left side: control plants showing normal, good growth and health (Pictures from PBGL, Joint FAO/IAEA Division b and Sokoine University in URT)

Fig. 2.4 Effects of heat stress application on seedlings. (**a**) After heat stress application some seedlings have survived and some have not. (**b**) All seedlings have died these lines are very sensitive to heat. (**c**) Only one to two seedling has survived and can be selected for multiplication. (**d**) Control (non-heat stress) healthy seedlings in hydroponics. (**e**) Heat stressed seedlings in hydroponics with injury to leaves some of them bad injured non tolerant to heat. (**f**) Control (non-heat applied) seedlings roots without damage. (**g**) After heat stress application, damage can be seen on roots of rice seedlings, bad injured leaves seedlings also have bad injured roots (Pictures from PBGL, Joint FAO/IAEA Division and Sokoine University, URT)

2.2.5 Assessment of Heat Tolerance at the Seedling Stage

Heat application to rice seedlings is illustrated in Fig. 2.5; visual symptoms of heat stress are: leaf yellowing, reduced leaf area, leaf tip burning, and entire leaf burning and leaf death. The biomass of seedlings is recorded as given Table 2.1 using shoot/root/whole seedlings weight (fresh and dry), root, and shoot length and seedling height data. The performance of test lines/cultivars/controls are compared to the standards (controls), e.g. heat tolerant cv. N22, moderate tolerant cv. IR64 and/or sensitive cultivars IR 52, and/or WAB 56 104. Most of lines showed poor growth or death at temperatures higher than 45 °C while some tolerant lines/cultivars showed better responses than sensitive. However tolerant lines showed slightly damaged compared to susceptible and highly susceptible lines, below scoring was used for classify genotypes/populations (Wei-hun et al. 2012).

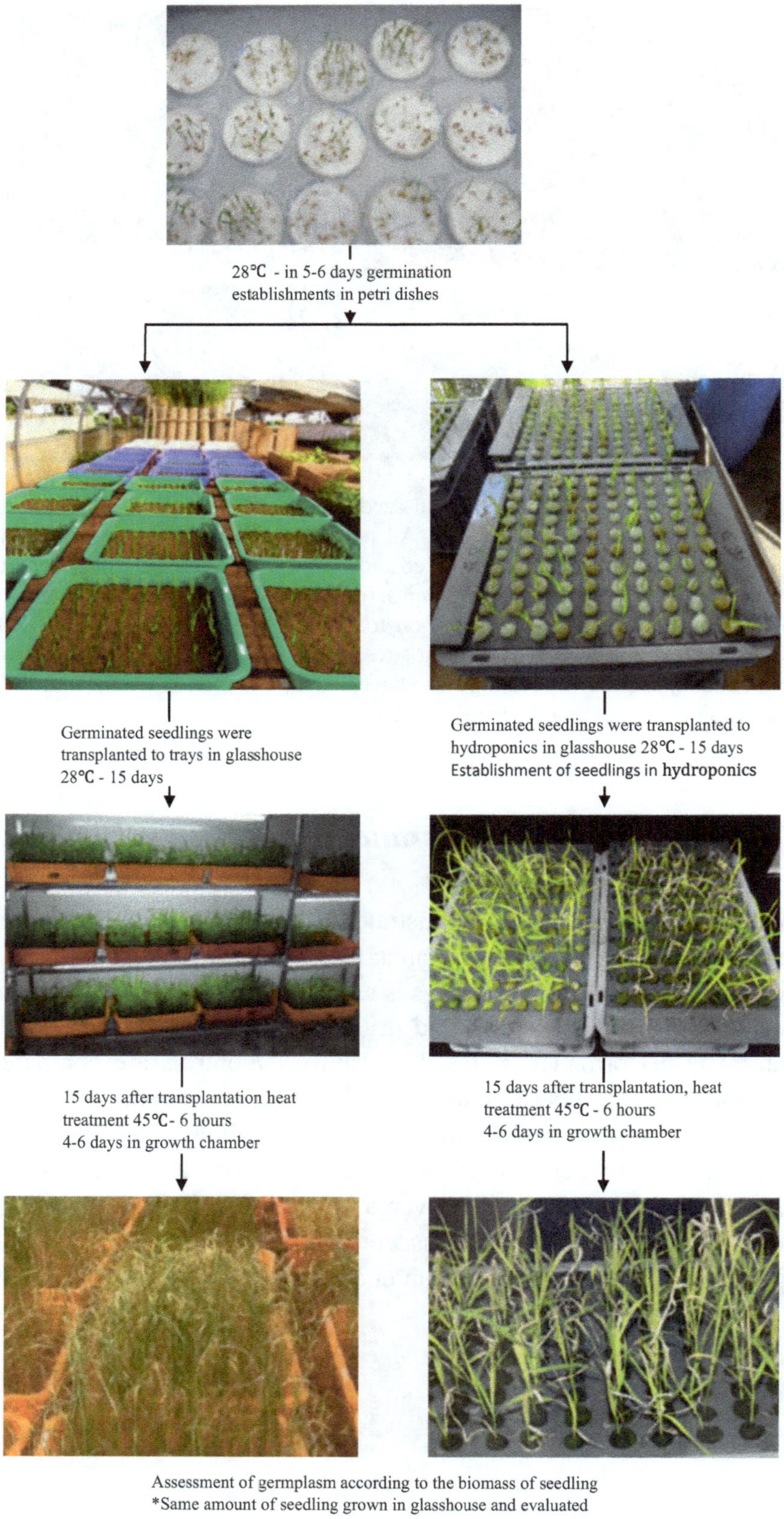

Fig. 2.5 Various steps of the protocol for screening rice seedlings for heat tolerance

Table 2.1 Evaluation scores of seedlings for heat tolerance

Score	Visual observation	Relative tolerance
1	Nearly normal growth very rarely leaf rolling and leaf tips	Tolerant
2	Most of leaves are rolled, yellowish and reduced leaves	Moderately tolerant
3	Most of the leaves are dry and some of them dies	Susceptible
4	Most of seedlings dying or already dead	Highly susceptible

2.2.6 Heat Tolerance Index (HTI)

After heat stress treatment, mutants were relocated to normal conditions with control plants. Two days after treatment data were recorded for shoot and root lengths, shoot and root fresh weights, shoot and root dry weights for stressed seedlings along with their respective controls grown under normal temperature.

A heat screening scale based on the response to stress was developed using parameters showing significant correlations with each other and with relative heat tolerance. Parameters were tested under control as well as heat stress and stress index was calculated for each parameter. The HTI is calculated by dividing the sum of individual scores for each parameter by the total score and multiplied by 100%. The genotypes are graded from tolerant to sensitive.

Stress tolerance indices (STIs) for the recorded traits are calculated as follows (Fernandez 1992; Blum 1988):

$$\text{STI} = (\text{Value under stress}/\text{Value at control}) \times 100$$

The genotypes were graded as tolerant, moderately tolerant, moderately sensitive and sensitive (Ashraf et al. 1999). Tables 2.2, 2.3, 2.4 and 2.5 summarize the data from one representative screening carried out on rice at the seedling stage in hydroponics and in pots and trays.

Table 2.2 Classification of heat tolerance in seedlings stage of various rice genotypes compared to medium heat tolerant cv. IR64 from NIAB Pakistan

More susceptible than control cv. IR64	Equivalent to control cv. IR64	More tolerant than control cv. IR64
Tested 159 mutant lines	HT-31, HT-51, HT-53, HT-81, HT-104, HT-114, HT-138	HT-18, HT-39, HT-92, HT-97, HT-98, HT-119

Table 2.3 Classification of heat tolerance at the seedling stage of various rice genotypes compared to heat tolerant (HT) cv. N22 from FAO/IAEA PBGL, Seibersdorf, Austria

More susceptible than control cv. N22	Equivalent to control cv. N22	More tolerant than control cv N22
LP-16, Super Basmati, IR 52, JC-1, JC-2, Saros 5, Sahel 108, Sım 2 Sunade, Supa, G 619, WAB 56 104, WAB 56 50, Pachaperumal, BG 300, BG 357, pokkali, bicol	8552, LP 7, HT-53, HT-104, HT-132, HT-138, HT-119IR 64, Sahel 317, HT-39,	8553, LP12, HT-97, HT-74

Table 2.4 Classification of heat tolerance at the seedling stage of various rice genotypes compared to heat tolerant (HT) cv. N22 from Jawaharlal Nehru University, India

More susceptible than control cv. N22	Equivalent to control cv. N22	More tolerant than control cv. N22
Tested 108 mutant lines	IR 64, D100/79 and 200 mutant lines	N22, D100/111 and D100/96

Table 2.5 Classification of heat tolerance at the seedling stage of 669 mutant lines compared to parental (wild type) and HT sensitive cv. WAB 56-104 from Sokoine University, UR Tanzania

More susceptible than control WAB 56-104 and parent	Equivalent to control cv WAB 56-104	More tolerant than control cv WAB 56-104
KR 28, KR 29, KR 30, CG14-20, CG14-52, CG14-53, CG14-65, CG14-77, CG14-78, CG14-80, CG14-79, CG14-81, WAB56-104-39, WAB56-104-40, WAB56-104-73, WAB56-104-74, WAB56-104-132, WAB56-104-133, WAB56-104-170, WAB56-104-171, WAB56-50-59, WAB56-50-60, WAB56-50-87, WAB56-50-88, WAB56-50-95, WAB56-50-96, WAB56-50-99, WAB56-50-124, WAB56-50-125, WAB56-50-143, WAB56-50-142, WAB56-50-133, WAB56-50-134 and WAB56-50-156	KR-9, KR-26, KR-39, KR-40, CG14-49, CG14-50, CG14-54, CG14-51, CG14-60, CG14-66, CG14-82, CG14-75, CG14-76, WAB56-104-12, WAB56-104-46, WAB56-104-70, WAB56-104-62, WAB56-104-136, WAB56-104-137, WAB56-104-119, WAB56-104-181, WAB56-50-70, WAB56-50-90, WAB56-50-91, WAB56-50-100, WAB56-50-107, WAB56-50-119, WAB56-50-139, WAB56-50-140, WAB56-50-146, WAB56-50-147	KR 3, KR 10, KR 27, KR 37, KR 38, CG14-5, CG14-6, CG14-7, CG14-13, CG14-16, CG14-21, CG14-22, CG14-59, CG14-61, CG14-62, CG14-63, CG14-64, WAB56-104-9, WAB56-104-18, WAB56-104-36, WAB56-104-43, WAB56-104-71, WAB56-104-76, WAB56-104-118, WAB56-104-141, WAB56-104-123, WAB56-104-150, WAB56-50-48, WAB56-50-51, WAB56-50-56, WAB56-50-82, WAB56-50-85, WAB56-50-97, WAB56-50-98, WAB56-50-123, WAB56-50-127, WAB56-50-138, WAB56-50-141, WAB56-50-152, WAB56-50-135

2.2.7 Recovery Stage After Heat Treatment

Heat tolerant lines that survive and recover after the heat stress at 45 °C may also be transplanted in hydroponics or soils. Selected seedlings can be grown on to maturity in hydroponic tanks filled with Yoshida solution changed every 2 weeks. Selected tolerant seedlings were gently teased out of the test tanks with care taken to keep roots intact. The base of the aerial part of each selected seedling is then wrapped in a sponge strip (10 × 2 × 1 cm) and the seedlings inserted into a recovery tank (Fig. 2.6a–c).

Heat tolerant lines may also be transplanted into 5 l pots containing soil. The seedlings were maintained individually and single plants and allowed to form tillers and panicles. Preliminary data comparisons may be recorded on better lines with respect to: plant height, number of tillers, days to flowering, number of panicles, panicle length, sterility and fertility of panicle as well as yield data such as seed number, weight, and thousand grain weights for each individual plant per pot. Seeds from single plants are kept separately for multiplication in the next generation (Fig. 2.6d).

Fig. 2.6 Recovery of seedlings after heat treatment and transferto hydroponics or pots. Selected seedlings transferred to hydroponics and pots (**a** and **c**) and pots (**c**). These lines can be grown on until maturity (**b** and **d**) (Pictures from Joint FAO/IAEA Division PBGL Lab, Sokoine University in URT and Nuclear Institute for Agriculture and Biology (NIAB), Pakistan)

2.3 Screening Protocol for Heat Tolerance in Rice at the Flowering Stage

2.3.1 Germination and Seedling Establishment in Hydroponics and Pots

Uniform healthy seedlings (8–10 days old) are selected and grown individually in hydroponics (Fig. 2.7). The glasshouse set up for rice: 28/20 °C day/night temperatures with 70% humidity and 16 h photoperiod. The hydroponics and pot culture set ups were the same as that described in Sects. 2.2 and 2.3. The glasshouse should be disease-free and well-lit by natural and/or artificial lighting. 15-cm pots are recommended for single plants as these give enough seed yield and panicle number for line evaluation.

Every 2 weeks, the hydroponics solution needs to be changed. Hydroponics is also aerated by agitation twice in a week. Pots and boxes with soil culture were irrigated twice a week to maintain soil saturation. N, P, K fertiliser applications were done prior to transplanting and panicle emergence, which may be variable depending on the genotype.

Fig. 2.7 (**a**) Pre-germinated rice seedlings in petri dishes, 8–10 day's old young seedlings transplanted to (**b**) pots and or (**c**) boxes (each box contains 9–10 plants). (**d**) Rice seedlings transplanted to hydroponics and grow on to the flowering stage (Pictures from Joint FAO/IAEA Division PBGL Lab)

Seedling establishment of control (standard) plant materials and test plant materials, care of seedlings in hydroponic, pots and boxes are same as described in Sects. 2.2 and 2.3.

2.3.2 *Heat Treatment at the Flowering Stage*

Two treatments were applied to test genotypes (1) plants with no heat stress exposure (control) and (2) plants subjected to heat stress at flowering stage. Breeding lines and cultivars are used, with standards N22, IR64 as a heat tolerant/moderately tolerant standards; IR 52 and/or WAB 56 100 were used as heat sensitive standards for comparative purposes. Plants were transferred to controlled environmental chambers/growth room conditions at the beginning of anthesis (first day) for heat stress treatments. Various heat stress treatments were investigated, all beginning at anthesis for various durations (2–6 h to 3–6 days) and at various temperatures ranging from 35 to 39 °C. Here we describe the most effective treatment, at 39 °C for 4 days, 6 h per day and 35 °C, 6 days, 6 h per day and 38 °C, 5 days and 4 h per day).

First Treatment A set of 12 mutant lines (one plant grown per pot) were taken for testing high temperature tolerance. The main tiller of each plant was tagged before exposing plants to high temperature at flowering. On the appearance of the first flower on the main tiller plants were transferred to growth chambers at 07:00 h before the anthesis process starts. Environmental control chambers are set at 39 °C with 80% relative humidity during the period of 6 hours 08:00–14:00 covering the anthesis period for each day. During anthesis, opened spikelets from four replicates (five spikelets from each plant) are marked with a permanent marker pen and moved to growth chamber for heat application.

Second Treatment Each genotype replicate consisted of ten pots with one to two plants per pot. Alternatively, 10 plants per genotype were grown up in soil filled boxes. Screens consisted of three replicates per genotype. Half (5) of the 10 pots or boxes were maintained in glasshouse conditions and served as controls. The other 5 were placed in a growth chamber and exposed to high temperatures at anthesis as described below. Illustration of heat stress treatments on rice at the flowering stage is given Fig. 2.8.

Heat stress treatment; At the beginning of anthesis, five pots per replicate of each genotype are transferred to growth chambers conditions in the morning 07:00 am with 80% humidity exposed to 35 °C for 6 days 6 h, and 38 °C 5 days 4 h respectively on consecutive days. The main tiller of each pot is tagged before exposing plants to high temperature at anthesis. At the first day of anthesis tagged plants were transferred to growth chambers set at 35–38 °C with 80% relative humidity as of from 7:00 AM every day. After each daily heat stress application, plants were returned to normal glasshouse conditions (30/20 °C day/night temperatures with 80% humidity). Plant phenotyping and heat stress treatment in growth chamber at flowering stage are shown in Fig. 2.9.

Fig. 2.8 Heat stress phenotyping in growth chambers at 35.0–39.0 °C/28 °C (day/night) for 6–4/18–20 h for 6–4 days 80% relative humidity (**a**, **c** and **d**). Heat stress application in growth chamber with pots, trays and hydroponics. (**b**) Enlarged view of panicles (Pictures from Jawaharlal Nehru University, India)

Fig. 2.9 Illustration of heat stress treatments on rice at the flowering stage

2.3.3 *Screening for Heat Tolerance at the Flowering Stage*

Spikelet fertility at maturity was used to screen heat tolerance of tested genotypes. Plant height and tiller number per plant were recorded. Mature panicles were harvested from five remaining pots for grain yield per plant recording, grain quality traits and calculating spikelet fertility (by counting empty and grain-filled spikelets). Main tillers were tagged and after harvest panicles were manually threshed, numbers of filled and unfilled grains per panicle were recorded. Each floret can be pressed between forefinger and thumb to determine if the grain is filled or not. Both partially and fully filled spikelets were categorized as filled spikelets.

Spikelet fertility is calculated as below;

$$\text{Spikelet fertility} = \frac{\text{Number of filled grains}}{\text{Total number of reproductive sites (floret)}} \times 100$$

Spikelet fertility was calculated in control and high temperature stress conditions. The negative effect (% decrease from control) was determined per genotype.

Reduced fertility of more than 65% is classed as highly susceptible, between 25–65% classed as intermediate (moderately susceptible) and less than 25% is classified as heat tolerant. This quantification has been adapted from species, ecotype and cultivar differences in spikelet fertility and harvest index of rice in response to high temperature stress (Prasad et al. 2006; Rang et al. 2011) (Tables 2.6, 2.7 and 2.8).

Table 2.6 Classification of heat tolerance at the seedling stage of various rice genotypes compared to the heat tolerant (HT) cv. N 22 from FAO/IAEA PBG Seibersdorf, Austria, (treatment 35 °C for 6 days, 6 h per days)

More susceptible than control cv. N22 Susceptible > 65% sterility	Equivalent to control cv. N22 Moderate tolerance: sterility from 25–65%	More tolerant than control cv. N22 Tolerant lines: sterility < 25%
LP-16, WAB 56 104, WAB 56-50	LP 7, IR 64, HT-39, HT-119,	8553, LP12, HT-97, II You 838, N 22, HT 74

Table 2.7 Classification of heat tolerance at the flowering stage of various rice genotypes from Jawaharlal Nehru University, India compared to the heat tolerant (HT) cv. N 22 (treatment 39 °C for 4 days, 4 h per day)

More susceptible than control cv. N22 Susceptible > 65% sterility	Equivalent to control cv. N22. Moderate tolerant sterility from 25–65%	More tolerant than control cv. N22 Tolerant line sterility < 25%
Tested 90 lines are susceptible than IR 64	IR 64 and D100/79	N22, D100/111 and D100/96

Table 2.8 Classification of heat tolerance at the flowering stage of mutant lines compare to parental (wild type) genotypes and HT sensitive cv. WAB 56-104 and parental lines from Sokoine University, UR Tanzania (treatment 38 °C for 5 days, 4 h per days)

More susceptible than control cv.WAB 56 Susceptible > 65% sterility	Equivalent to control cv. WAB 56-104. Moderate tolerant sterility from 25–65%	More tolerant than to control cv WAB 56-104 Tolerant lines sterility < 25%
KR-3; KR-38; CG14-59; CG14-63; CG14-64; WAB56-104-18; WAB56-104-118; WAB56-104-123; WAB56-50-85; WAB56-50-98; WAB56-50-141; WAB56-50-152	KR-27; CG14-5; CG14-6; CG14-7; CG14-61; CG14-62; WAB56-104-141; WAB56-50-51; WAB56-50-97; WAB56-50-123; WAB56-50-127; WAB56-50-135	KR-37; CG14-13; CG14-16; CG14-21; CG14-22; WAB56-104-9; WAB56-104-36; WAB56-104-43; WAB56-104-71; WAB56-104-76; WAB56-104-150; WAB56-50-48; WAB56-50-96; WAB56-50-82; WAB56-50-138

Chapter 3
Validation of Screening Protocols for Heat Tolerance in Rice

3.1 Background Analysis

An alternative to screening for rice responses to heat at seedling and reproductive discussed in Chap. 2. This chapter covers further validation of identified lines through: (1) physiological and biochemical indicators that are reported to be associated with plant stress responses to classify the mechanism of heat tolerance of screened lines/varieties; (2) Field trials under real stress conditions provide conclusive data for screening and selection for advancing promising lines to confirm the methods developed in Chap. 2. Yield data were used to assess tested rice genotypes and comparisons made to heat tolerant/susceptible standards. Tested lines were classified as sensitive, intermediate and tolerant.

Biochemical changes are another aspect of plant acclimation to heat stress and heat stress may also affect leaf chlorophyll content and membrane stability through lipid peroxidation, leading to the production of peroxide ions and MDA (Wahid et al. 2007). Increasing the temperature changes of 37 °C/30 °C (day/night) increase the electrolyte leakage percentage and MDA content in rice (Zhang et al. 2009; Liu et al. 2013).

Identified lines/varieties were subjected to physiological and biochemical tests, such as electrolyte leakage, MDA level, total protein content, antioxidant enzymes (catalase, ascorbate peroxidase and superoxide dismutase) activity at seedling, vegetative stages and flowering stages. Results of the physiological and biochemical measurements confirmed heat tolerance behaviour of selected germplasm. This work was carried out at the Stress Physiology and Molecular Biology Laboratory of School of Life Sciences, Jawaharlal Nehru University New Delhi, India.

After screening at the seedling stage and/or flowering stages, the identified heat tolerant genotypes were tested in field conditions to evaluate real performance of selected promising lines in hot spot areas at various locations (in Cuba, Pakistan and UR Tanzania), known to experience heat stress. Field trials usually involve replicated plots that take up a lot of space and often in more than one location. Therefore,

F. Sarsu et al., *Pre-Field Screening Protocols for Heat-Tolerant Mutants in Rice*,
https://doi.org/10.1007/978-3-319-77338-4_3

breeders need to be selective in what is to be tested. The pre-field screening protocols described in Chap. 2 provide potential candidates that may be advanced for field evaluations thus saving time, space and costs on lines with no potential.

3.2 Validation of Physiological and Biochemical Indicators

3.2.1 Physiological and Biochemical Characterization of Heat Tolerant Lines at the Seedling Stage

Physiological and biochemical changes to heat stress in rice include electrolyte leakage, lipid peroxidation level, MDA content and antioxidant enzyme activity (Campos et al. 2003; Heath and Packer 1986; Bajji et al. 2001; Nakano and Asada 1981; Oberley and Spitz 1985). Selected heat tolerant mutant lines of rice recorded a low level of electrolyte leakage thereby indicating less damage to membrane, as compared to the wild type parent medium heat tolerant IR64 (Fig. 3.1). Moreover MDA content (an indicator of lipid peroxidation) was assessed as a test for heat tolerance. It was observed that heat tolerant mutant lines showed relatively low MDA levels (Fig. 3.2). MDA is a reactive aldehyde and is one of the many reactive electrophile species that cause toxic stress in cells. Thus, the lower the level of MDA shows better plant performance. Antioxidant enzymes such as CAT and APX (ascorbate peroxidase) showed higher activity on the first day of heat stress treatment and SOD showed higher activities on the third day of heat stress treatment, as compared to wild type genotypes (Fig. 3.3) indicating a heat stress tolerant physiology.

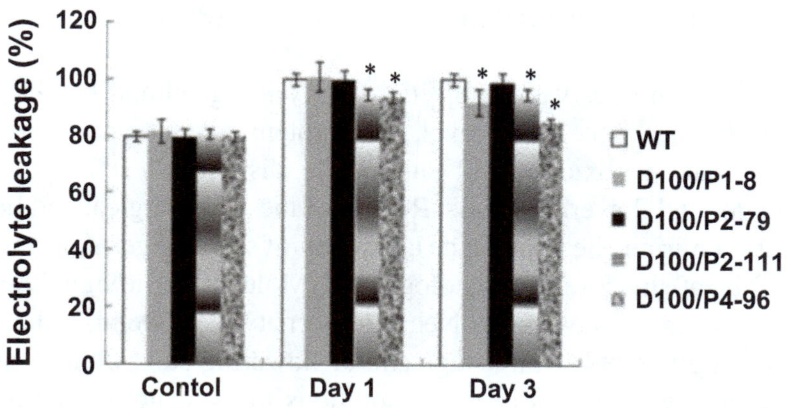

Fig. 3.1 Percentage electrolyte leakage in putative heat tolerant mutant lines as compared to IR64. Electrolyte leakage of IR64 under stress was taken as 100% and used as a reference to compare mutant lines. Day 1 and Day 3 indicate 1st day and 3rd day after heat stress treatment. The data represent means ± SE of three biological replicates. Bars with stars are statistically significant ($p < 0.05$)

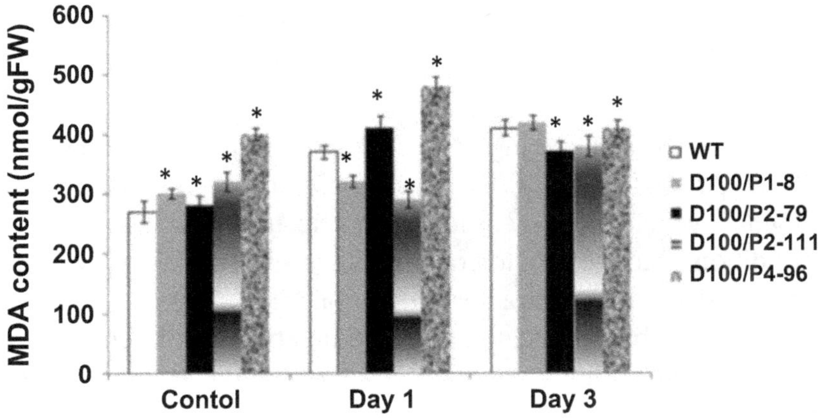

Fig. 3.2 MDA content in putative heat tolerant mutant lines compared to IR64. Day 1 and Day 3 indicate 1st day and 3rd day after heat stress treatment. The data represent means ± SE of three biological replicates. Bars with stars are statistically significant ($p < 0.05$)

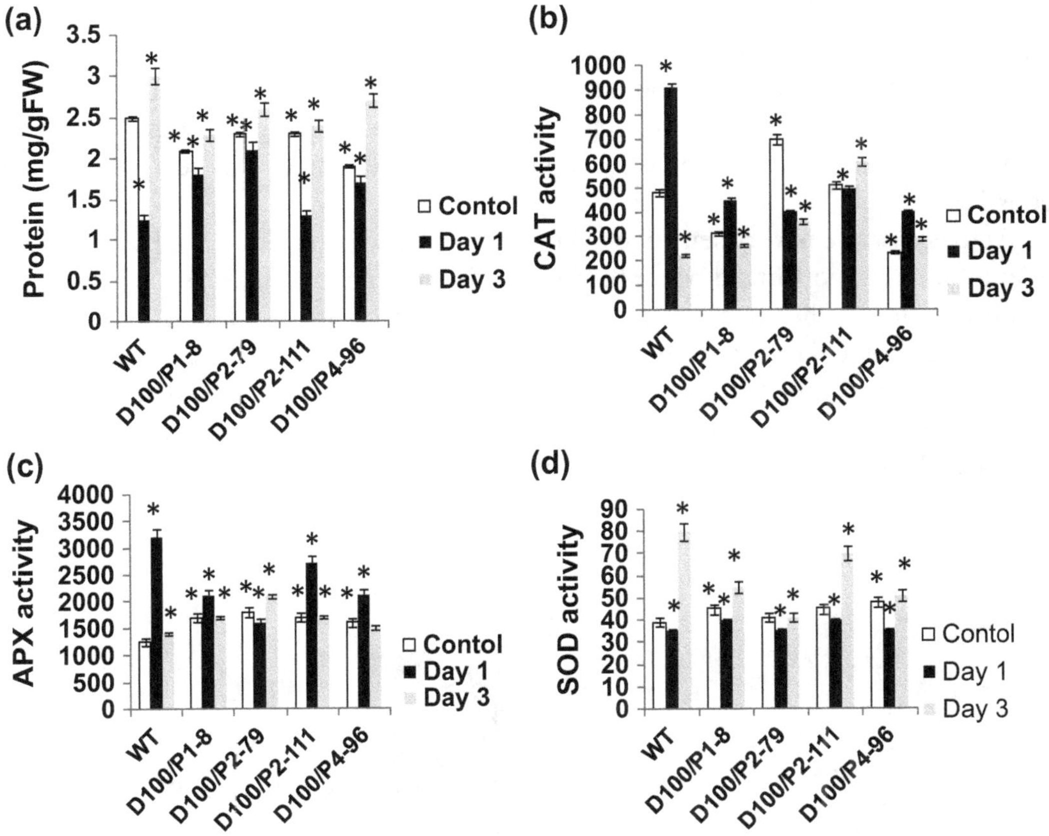

Fig. 3.3 Total protein content, CAT, APX and SOD activity in putative heat tolerant mutant lines as compared to wild type (IR64). Day 1 and Day 3 indicate 1st day and 3rd day after heat stress treatment. The data represent means ± SE of three biological replicates. Bars with stars are statistically significant ($p < 0.05$)

3.2.2 *Physiological and Biochemical Characterization of Heat Tolerant Lines at the Flowering Stage*

Before heat treatments at flowering of the putative heat tolerant lines (D100/P1-8, D100/P1-79, D100/P2-111, D100/P4-96), an assessment was made for membrane stability at the late vegetative stage. Electrolyte leakage was similar in all genotypes in control conditions, but the heat tolerant lines had superior electrolyte leakage than the wild type standard under heat stress conditions (Fig. 3.4).

Morphological and biochemical analysis of putative heat tolerant mutant plants were analysed at the flowering stage as this is a temperature sensitive stage (Jagadish et al. 2007). The mutants phenotype were analysed under optimum temperatures in glasshouse conditions immediately after heat stress treatment (40 °C for 4 h) and after recovery (72 h, 28 °C). The putative heat tolerant mutants performed better than their wild type parent (Fig. 3.5). The mutant plants were healthy and continued to grow after the heat stress treatment, whereas a reduction in growth was observed in wild type plants after heat stress and during the recovery period. Data on panicle length, height, number of tillers and number of seeds per panicle were found to be superior in the heat tolerant mutants than for wild type (Fig. 3.5). The putative mutant lines also produced more vegetative and reproductive growth under normal as well as heat stress conditions, indicating that tolerance to heat stress was constitutive.

Anthesis in rice is very sensitive to heat stress (Jagadish et al. 2007). Putative heat tolerant mutants performed better than their wild type parent under heat stress at the flowering stage. Phenotypic data were recorded in control optimum temperature conditions, immediately after heat stress treatment (40 °C for 4 h) for three subsequent days (to span main panicle flowering) and after recovery (72 h 28 °C). Besides healthier appearance and better growth, the putative heat tolerant mutants recorded higher spikelet fertility and yield components such as panicle length, number of

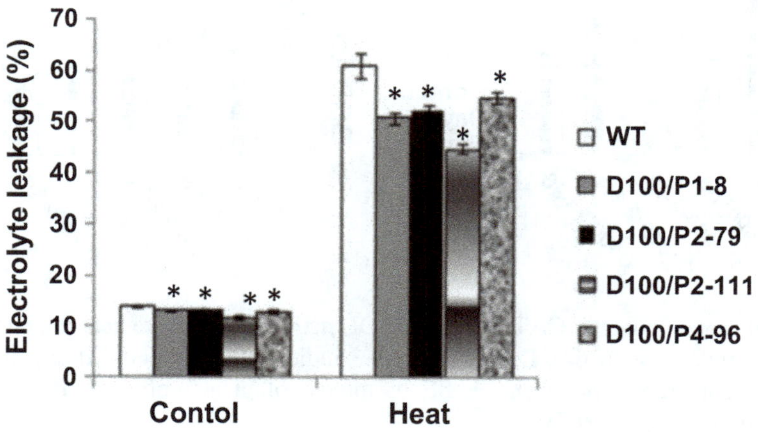

Fig. 3.4 Electrolyte leakage (%) of putative heat tolerant lines at pre-flowering stage. The data represent means ± SE of three biological replicates. Bars with stars are statistically significant ($p < 0.05$)

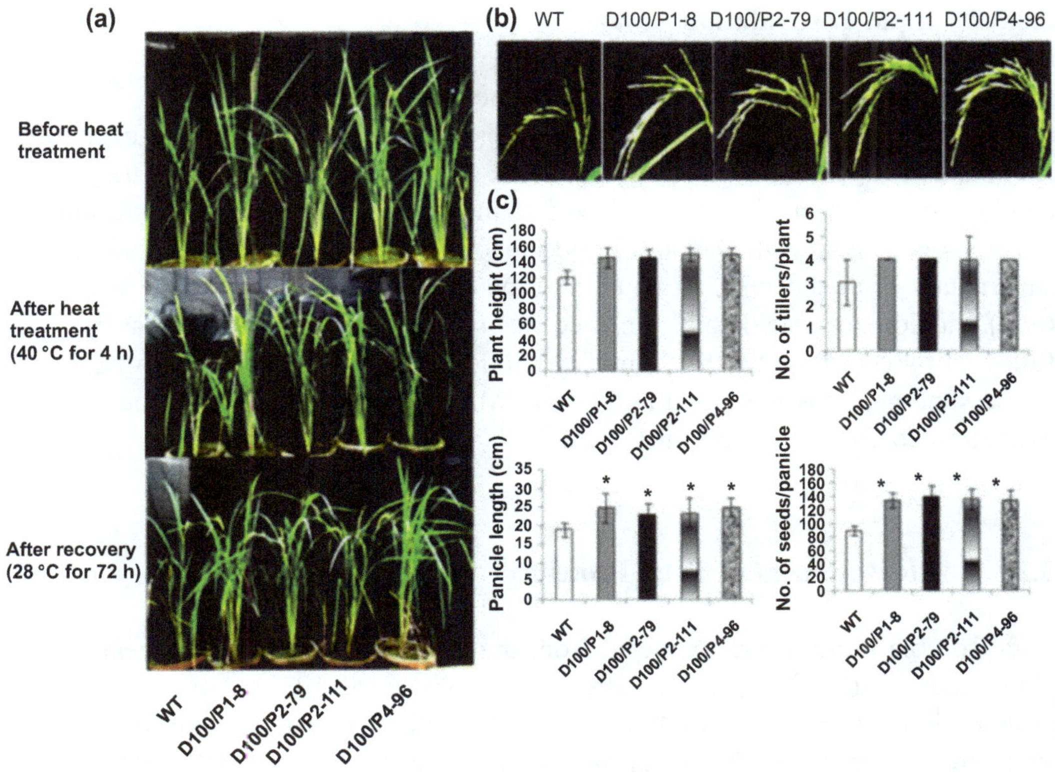

Fig. 3.5 Morphology and morphological parameter analysis of heat tolerant mutants. (**a**) Wild type and heat tolerant mutants before stress, immediately after stress and during the recovery phase. (**b**) Panicle phenotype of wild type and heat tolerant mutants after recovery. (**c**) Different morphological parameters in wild type and heat tolerant mutants after recovery. The data represent means ± SE of three biological replicates. Bars with stars are statistically significant ($p < 0.05$)

tillers, and number of spikelets per panicle grain yield and plant height per plot as compared to wild type. The mutant lines had a higher vegetative and reproductive growth under both normal and heat stress conditions, suggesting that the genetic controls were constitutive.

3.2.3 *Methods*

3.2.3.1 Electrolyte Leakage Measurement

Analysis of electrolyte leakage is carried out following the protocol of Bajji et al. (2002). Leaf samples from control and heat-treated plants are harvested and washed with distilled water to remove any surface adhering ions. About 100 mg of the tissue are dipped immediately into 20 ml of de-ionized water. After incubating the leaf tissues at 37 °C for 2 h, the electrical conductivity (E1) of the immersion solution is measured using a conductivity meter (ELEINS, Inc., India). To determine the total conductivity (E2), the samples with immersion solution (effusate) are autoclaved for 15 min at 121 °C and the conductivity of the effusate is measured after cooling it to room temperature. Relative electrical conductivity is measured by the formula: percentage of electrolyte leakage = E1/E2 × 100.

3.2.3.2 Lipid Peroxidation Assay

Lipid peroxidation is estimated measuring the formation of MDA. MDA content is quantified by thiobar-bituric acid reactive substances assay (Heath and Packer 1986; Larkindale and Knight 2002). For this, about 100 mg of leaf tissue from the control and heat-treated plants is homogenized in 5 mL of 5% (w/v) trichloroaceticacid and the homogenate centrifuged at 12,000×*g* for 10 min at room temperature. The supernatant is mixed with an equal volume of thiobarbituric acid [0.5% in 20% (w/v) trichloroaceticacid] and the mixture is then boiled for 25 min at 100 °C, followed by centrifugation for 5 min at 7500 g to clarify the solution. Absorbance of the supernatant is measured at 532 nm. MDA equivalents are calculated by the extinction coefficient of 155 mM^{-1} cm^{-1}.

3.2.3.3 Measurement of Total Proteins

Total soluble proteins are extracted from the rice leaves (100 mg) of heat treated plants at the seedling stage using Zivy's buffer (Zivy et al. 1983). Amount of soluble proteins is estimated by Bradford's assay (Bradford 1976) with the help of the standard curve prepared using various known concentrations of other protein, such as Bovine Serum Albumin (BSA fraction V), prepared by using different concentration of Bovine Serum Albumin (Ernst and Zor 2010).

3.2.3.4 Antioxidant Enzyme Activity Assay

About 100 mg leaf material from the heat treated plants was homogenized in ice-cold 50 mM K-PO_4 buffer (pH 7.5) containing 2 mM EDTA and 0.1 mM phenylmethylsulphonylfloride (PMSF). The homogenizing buffer for APX additionally contained 2 mM of Na-ascorbate. The homogenate was centrifuged at 12,000×*g* for 10 min at 4 °C and the supernatant was used for enzyme assay. Total protein content in supernatant was determined following the method of Bradford (1976).The activity of SOD was measured using the method of Giannopolitis and Reis (1977). The assay mixture (1 mL) for SOD contained 79.2 mM Tris–HCI buffer (pH 8.9), containing 0.12 mM EDTA, 10.8 mM tetra ethylene diamine, bovineserum albumin (3%), 6 mM nitroblue tetrazolium (NBT), 60 μM riboflavin in 5 mM KOH and 5 μg of enzyme extract. The increase in absorbance due to formazan formation was read at 560 nm. The increase in absorbance in the absence of enzyme was taken as 100%, and 50% initial was taken as equivalent to 1 unit of SOD activity. APX activity was determined according to Nakano and Asada (1981). The reaction mixture in a total volume of 1 ml consisted of 50 mM (pH 7.5) K-PO4buffer, 0.1 mM EDTA, 0.25 mM ascorbate, 10 mM H_2O_2 and enzyme extract. H_2O_2-dependent oxidation of ascorbate was followed spectrophotometrically by recording the decrease in absorbance at 290 nm (D = 2.8 $mM^{-1}cm^{-1}$). Slope values of absorbance in 290 nm are considered for rate calculation. Catalase (CAT) activity was determined following the method of Aebi (1983) by measuring the decrease in absorbance at 240 nm due to the decomposition of H_2O_2 ($D = 40\ M^{-1}cm^{-1}$). The slope value of the rapid decline in 240 nm absorption was considered for rate calculation. The reaction mixture in 1 ml contained 50 mM K-PO_4 buffer (pH 7.0) with leaf extract equivalent to 4–5 μg protein. The reaction was initiated by adding H_2O_2 to a final concentration of 20 mM. All the experiments involved in physiological and biochemical analyses were repeated three times using three independent mutant plants from each line.

3.3 Validation Protocols of Rice Heat Tolerance Under Field Conditions

After screening at the seedling stage and flowering stages (as described in Chap. 2) identified heat tolerant genotypes were tested in field conditions to validate the pre-field screening protocols in hotspot test locations.

The field trials were conducted in different places which were known for heat stress in Cuba, Pakistan and UR Tanzania. Field experiments were planted in completely randomized block designs with three or four replications per location. To further verify the results of seedling and reproductive stages screening protocols data were collected on:

- Time of day for flowering (anthesis)
- Plant height
- Number of tillers per plant
- Panicle length
- Spikelet number of the main tiller panicle
- Spikelet fertility (ratio of filled/unfilled grains)
- Thousand kernel weight
- Paddy yield

Data were recorded for plant height; number of productive tillers per plant, panicle length, and spikelet per main panicle, panicle fertility, thousand seed weight and yield were recorded. From these data the tested genotypes including mutant lines were categorized as below (Tables 3.1, 3.2 and 3.3).

Screened lines were evaluated for various agro-morphological characters in field tests and compared to standards including local cultivars and parental lines at different locations and years. Temperature and rainfall of the locations were recorded and compared with long term temperature means. Comparisons were made for yield performance as well as yield components such as spikelet number, thousand kernel weight and filled/unfilled grain ratio under natural high temperature conditions. The results of field screening confirmed the results of the laboratory (physiological and biochemical) testing and also pre-field screening results at both the seedling and flowering stage.

Table 3.1 Classification of heat tolerance of rice genotypes in field conditions at NIAB, Pakistan compared to medium heat tolerant standard cv. IR64

More susceptible than control cv. IR64	Equivalent to control cv. IR64	More tolerant than control cv. IR64
HS cv. 'Super Basmati', tested 30 mutant lines	IR 64, HT-18, HT-29, HT-31, HT-39, HT-51, HT-53, HT-98, HT-104, HT-114, HT-119, HT-132 and HT-138	HT-74, HT-81, HT-92 and HT-97

Table 3.2 Classification of heat tolerance rice genotypes in field conditions at INCA, Cuba compared to the control parent and local cultivars

More susceptible than control parent and local cultivars	Equivalent to control parent and local cultivars	Tolerant/better than control parent and local cultivars
LP-16, A 82	8551, 8554, 8555, LP-10, LP 9, LP-7, LP 8,	8553,LP12,8552,8555 Guillemar[a]

[a]Guillemar cv. released as heat tolerant variety in 2015 in Cuba

Table 3.3 Classification of heat tolerance of rice genotypes compare to parental (wild type) genotypes and HT sensitive cv. WAB 56-104 under field conditions from Sokoine University, URT Tanzania

More susceptible than control WAB 56- and parent	Equivalent to control cv WAB 56-104	More tolerant than control cv WAB 56-104
KR, WAB56-50, WAB56-104, WAB56-50-104-36-1, KR 38-1 WAB56-50-127-3, WAB56-50-98-1, KR 27-1	KR-27, WAB56-50-51, WAB56-50-97, WAB56-50-123, WAB56-50-127, WAB56-50-135, KR-37, WAB56-104-9	WAB56-50-85-3, WAB56-50-82-1, WAB56-50-74-1, WAB56-50-56-1, WAB56-104-36-1, WAB56-104-141-1, WAB56-104-141-3, WAB56-50-127-3, WAB6-104-36-1, WAB56-104-141-2, WAB56-50-97-3

Chapter 4
Conclusion

In conclusion, the comparison of candidate mutant rice lines with controls (standards) using various physiological and biochemical indicators showed that the selected heat tolerant mutant lines exhibited less electrolyte leakage and reduced levels of MDA, both indicative of improved plant performance under stress.

These results indicated that response of rice to heat stress tolerance could be assessed at the seedling and reproductive stages. There is a close correlation between the screening results of seedlings and plants at reproductive stages in the field and the laboratory—physiological and biochemical—of these genotypes, thus validating that the protocols developed can reliably identify heat stress tolerant genotypes in a mutation breeding program. Moreover, the field data showed good correlations with the pre-field screening results at both the seedling and flowering stages, with ~70% of the heat tolerant genotypes screened during the pre-field tests showing improved heat tolerance in the field. Thus, reducing the numbers from a few thousands to less than 100 potential individuals or lines that can be tested further in the field where they can be validated under more rigorous field conditions using replicated row or plot trials. This important finding indicates that heat tolerant genotypes may be selected in the green house/growth chamber conditions prior to field trials.

Heat stress tolerance may involve different mechanisms. These protocols are primarily designed to be used by rice breeders who need practical high-throughput screens to process large populations such as mutant populations, but also segregating populations, advanced generations and germplasm collections for heat stress tolerance. The techniques described here can facilitate the identification of different sources of heat stress tolerance offering opportunities for pyramiding traits and developing high yielding heat tolerant rice varieties. Thereby developed protocols allow plant breeders to screen large rice germplasms in growth chamber conditions

F. Sarsu et al., *Pre-Field Screening Protocols for Heat-Tolerant Mutants in Rice*,
https://doi.org/10.1007/978-3-319-77338-4_4

at the seedling as well as reproductive stage to identify the potential resilience of rice mutant population/lines to heat stress and to define phenotypic and physiological traits that are associated with heat stress adaptation.

Correction to: Pre-Field Screening Protocols for Heat-Tolerant Mutants in Rice

Correction to:
F. Sarsu et al., *Pre-Field Screening Protocols for Heat-Tolerant Mutants in Rice*,
https://doi.org/10.1007/978-3-319-77338-4

The book was inadvertently published with several mistakes, which have now been corrected.

Copyright page: The affiliations of the authors Brian P. Forster and Ivan Ingelbrecht have been corrected to:

Brian P. Forster
Plant Breeding and Genetics Laboratory,
Joint FAO/IAEA Division
International Atomic Energy Agency
Vienna, Austria

Ivan Ingelbrecht
Plant Breeding and Genetics Laboratory,
Joint FAO/IAEA Division
International Atomic Energy Agency
Vienna, Austria

Preface: The second author name 'Priyanka Das' has been deleted.

The updated online version of the chapter can be found at
https://doi.org/10.1007/978-3-319-77338-4
https://doi.org/10.1007/978-3-319-77338-4_2

F. Sarsu et al., *Pre-Field Screening Protocols for Heat-Tolerant Mutants in Rice*,
https://doi.org/10.1007/978-3-319-77338-4_5

Chapter 2: The legends to Figure 2.8 and Figure 2.9 have been interchanged. The correct legends are:

Fig. 2.8 Heat stress phenotyping in growth chambers at 35.0–39.0 °C/28 °C (day/night) for 6–4/18–20 h for 6–4 days 80% relative humidity (**a**, **c** and **d**). Heat stress application in growth chamber with pots, trays and hydroponics. (**b**) Enlarged view of panicles (Pictures from Jawaharlal Nehru University, India)

Fig. 2.9 Illustration of heat stress treatments on rice at the flowering stage

Figure 2.8 has been replaced with a new version.

References

Aebi H (1983) Catalase. In: Bergmeyer HU (ed) Methods of enzymatic analysis. Academic Press, New York, pp 273–288

Ashraf M, Khan AH, Azmi AR, Naqvi SSM (1999) Comparison of screening techniques used in breeding for drought tolerance in wheat. In: SSM N (ed) Proceedings of the "new genetical approaches to crop improvement" -II. Karach Printing Press, Karachi, pp 513–525

Bado S, Forster BP, Mukhtar AGA, Cieslak JJ, Berthold G, Luxiang L (2016) Protocols for pre-field screening of mutants for salt tolerance in rice, wheat and barley. Springer open access, ISBN 978-3-319-26590-2 (eBook). http://www.springer.com/us/book/9783319265889

Bahuguna RN, Jagadish KS (2015) Temperature regulation of plant phenological development. Environ Exp Bot 111:83–90

Bahuguna RN, Pal JJ, Shah D, Lawas LM, Khetarpal S, Jagadish KS (2015) Physiological and biochemical characterization of NERICA rice: a novel source of heat tolerance at the vegetative and reproductive stages in rice. Physiol Plant 154(4):543–559

Bajji M, Kinet JM, Lutts S (2001) The use of the electrolyte leakage method for assessing cell membrane stability as a water stress tolerance test in durum wheat. Plant Growth Regul 00:1–10

Bajji M, Kinet JM, Lutts S (2002) The use of the electrolyte leakage method for assessing cell membrane stability as a water stress tolerance test in durum wheat. Plant Growth Regul 36:61–70

Baker JT, Allen LH, Boote KJ (1992) Temperature effects on rice at elevated CO2 concentration. J Exp Bot 43:959–964

Barakat MM, Al-Doss AA, Elshafei AA, Moustafa KA (2011) Identification of new microsatellite marker linked to the grain filling rate as indicator for heat tolerance genes in F2 wheat population. AJCS 5(2):104–110. ISSN:1835-2707

Barnabas B, Jager K, Feher A (2008) The effect of drought and heat stress on reproductive processes in cereals. Plant Cell Environ 31:11–38

Blum A (1988) Plant breeding for stress environments. CRC Press, Boca Raton, FL, p 212

Bradford MM (1976) A rapid and sensitive method for the quantitation of microgram quantities of protein utilizing the principle of protein-dye binding. Anal Biochem 72:248–254

Campos PS, Quartin V, Ramalho JC, Nunes MA (2003) Electrolyte leakage and lipid degradation account for cold sensitivity in leaves of Coffea sp. Plants. J Plant Physiol 160:283–292

Ernst O, Zor T (2010) Linearization of the Bradford protein assay. J Visual Expt. https://doi.org/10.3791/1918

Driedonks N, Rieu I, Vrizen WH (2016) Breeding for plant heat tolerance at vegetative and reproductive stages. Plant Reprod 29:67. https://www.ncbi.nlm.nih.gov/pmc/articles/PMC4909801/

F. Sarsu et al., *Pre-Field Screening Protocols for Heat-Tolerant Mutants in Rice*,
https://doi.org/10.1007/978-3-319-77338-4

Fernandez GCJ (1992) Effective selection criteria for assessing plant stress tolerance. In: Proceedings of the international symposium on adaptation of vegetables and other food crops in temperature and water stress, August 13–16, Shanhua, Taiwan, pp 257–270

Forster BP, Till BJ, Ghanim AMA, Huynh HO, Burstmayr H, Caligari PD (2014) Accelerated plant breeding. CAB Rev 43:1–16

Giannopolitis CN, Reis SK (1977) Superoxide dismutase I occurrence in higher plants. Plant Physiol 59:309–314

Hasanuzzaman M, Nahar K, Fujita M (2013) Extreme temperatures, oxidative stress and antioxidant defense in plants. In: Vahdati K, Leslie C (eds) Abiotic stress-plant responses and applications in agriculture. Tech Rijeka, Croatia, pp 169–205

Heath RL, Packer L (1986) Photoperoxidarion in isolated chloroplasts. I. Kinetics and stoichiometry of fatty acid peroxidation. Arch Biochem Biophys 125:189–198

IAEA (2017) Mutant variety database. https://nucleus.iaea.org/Pages/mvd.aspx

IPCC (2012) Managing the risks of extreme events and disasters to advance climate change adaptation. https://www.ipcc.ch/pdf/special-reports/srex/SREX_Full_Report.pdf

Jagadish SVK, Craufurd PQ, Wheler TR (2007) High temperature stress and spikelet fertility in rice (*Oryza sativa* L.) J Exp Bot 58:1627–1635

Jagadish SVK, Muthurajan R, Oane R, Wheeler TR, Heuer S, Bennett SJ, Craufurd PQ (2010) Physiological and proteomic approaches to address heat tolerance during anthesis in rice (*Oryza sativa* L.) J Exp Bot 61:143–156

Jagadish SVK, Craufurd P, Shi W, Oane R (2014) A phenotypic marker for quantifying heat stress impact during microsporogenesis in rice (*Oryza sativa* L.) Funct Plant Biol 41:48–55

Janska A, Marsik P, Zelenkova S, Ovesna J (2010) Cold stress and acclimation: what is important for metabolic adjustment. Plant Biol 12:395–405

Kondamudi R, Swamy KN, Chakravarthy DVN, Vishnuprasanth V, Rao YV, Rao PR, Sarla N, Subrahmanyam D, Voleti SR (2012) Heat stress in rice – physiological mechanisms and adaptation strategies page under the book crop stress and its management perspective and strategies. In: Venkateswarlu B, Shanker AK, Maheswari M (eds) Crop stress and its management: perspectives and strategies. Springer, Dordrecht, pp 193–223

Krasensky J, Jonak C (2012) Drought, salt, and temperature stress-induced metabolic rearrangements and regulatory networks. J Exp Bot 63(4):1593–1608. https://doi.org/10.1093/jxb/err460

Kumari S, Roy S, Singh P, Singla-Pareek SL, Pareek A (2013) Cyclophilins: proteins in search of function. Plant Signal Behav 8:e22734

Larkindale J, Knight MR (2002) Protection against heat stress-induced oxidative damage in Arabidopsis involves calcium, abscisic acid, ethylene, and salicylic acid. Plant Physiol 128:682–695

Li CY, Peng CH, Zhao QB, Xie P, Chen W (2004) Characteristic analysis of the abnormal high temperature in 2003 midsummer in Wuhan City. J Cent Chin Norm Univ 38:379–381

Liu QH, Wu X, Li T, Ma JQ, Zhou XB (2013) Effects of elevated air temperature on physiological characteristics of flag leaves and grain yield in rice. Chil J Agric Res 73:85–89

Manigbas N, Lambio LA, Madrid LB, Cardenas CC (2014) Germplasm innovation of heat tolerance rice for irrigated conditions in the Philippines. Rice Sci 21:162–169

Mason RE, Mondal S, Beecher FW, Pacheco A, Jampala B, Ibrahim ANH (2010) QTL associated with heat susceptibility index in wheat (*Triticum aestivum* L.) under short-term reproductive stage heat stress. Euphytica 174:423–436. https://doi.org/10.1007/s10681-010-0151

Matsui T, Omasa K, Horie T (1997) High temperature induced spikelet sterility of japonica rice at flowering in relation to air humidity and wind velocity conditions. Jpn J Crop Sci 66:449–455

Matsushima S, Ikewada H, Maeda A, Honda S, Niki H (1982) Studies on rice cultivation in the tropics, yielding and ripening responses of the rice plant to the extremely hot and dry climate in Sudan. Jpn J Trop Agric 26:19–25

Mittler R, Finka A, Goloubinoff P (2012) How do plants feel the heat? Trends Biochem Sci 37:118–125

Moreno AA, Orellana A (2011) The physiological role of the unfolded protein response in plants. Biol Res 44:75–80

Mottaleb KA, Rejesus RM, Mohanty S, Murty MVR, Li T, Valera HG, Gumma MK (2012) Ex ante assessment of a combined drought- and submergence-tolerant rice variety in the presence of climate change. In: Paper presented at the Agricultural & Applied Economics Association annual meeting, Seattle, Washington, 12–14 August, IRRI, Philipinnes

Munns R, Tester M (2008) Mechanisms of salinity tolerance. Annu Rev Plant Biol 59:651–681

Nakano Y, Asada K (1981) Hydrogen peroxide is scavenged by ascorbate-specific peroxidase in spinach chloroplasts. Plant Cell Physiol 22:867–880

Oberley LW, Spitz DR (1985) Assay of superoxide dismutase using nitroblue tetrazolium. In: Greenwald RA (ed) Handbook of methods for oxygen radical research. CRC Press, Boca Raton, FL, pp 217–227

Osada A, Sasiprapa V, Rahong M, Dhammanuvong S, Chakrabandho H (1973) Abnormal occurrence of empty grains of indica rice plants in the dry hot season in Thailand. Proc Crop Sci Soc Jpn 42:103–109

Pareek A, Singla SL, Grover A (1995) Immunological evidence for accumulation of two high molecular weight (104 and 90 kDa) HSPs in response to different stresses in rice and in response to high temperature stress in diverse plant genera. Plant Mol Biol 29:293–301

Pareek A, Singla SL, Grover A (1997) Short-term salinity and high temperature stress-associated ultrastructural alterations in young leaf cells of *Oryza sativa* L. Ann Bot 80:629–639

Pareek A, Sopory SK, Bohnert H, Govindjee J (2010) Abiotic stress adaptation in plants: physiological, molecular and genomic foundation. Springer, Dordrecht, ISBN: 978-90-481-3111-2

Poli Y, Basava RK, Panigrahy M, Vinukonda VP, Dokula NR, Voleti SR (2013) Characterization of a Nagina22 rice mutant for heat tolerance and mapping of yield traits. Rice 6:36. https://doi.org/10.1186/1939-8433-6-36

Prasad PVV, Bootee KJ, Sheehy JE, Thomas JMG (2006) Species, ecotype and cultivar differences in spikelet fertility and harvest index of rice in response to high temperature stress. Field Crop Res 95:398–411

Prasanth VV, Chakravarthi DVN, Vishnu KT, Venkateswara RY, Panigrahy M, Mangrauthia SK (2012) Evaluation of rice germplasm and introgression lines for heat tolerance. Ann Biol Res 3:5060–5068

Rang ZV, Jagadish SVK, Zhou QM, Crafurd PQ, Heuer S (2011) Effect of high temperature and water stress on pollen germination and spikelet fertility in rice. Environ Exp Bot 70(1):58–65

Ruelland E, Zachowski A (2010) How plants sense temperature. Environ Exp Bot 69:225–232

Rutger JN (2009) The induced sd-1 mutant and other useful mutant genes in modern rice varieties. In: Shu QY (ed) Induced plant mutations in the genomics era. In: Proceeding of an international joint FAO/IAEA symposium, Rome, pp 44–47

Sailaja BD, Subrahmanyam S, Neelamraju T, Vishnukiran YV, Rao P, Vijalakshmi SR, Voleti VP, Mangrauthia K (2015) Integrated physiological, biochemical, and molecular analysis identifies important traits and mechanisms associated with differential response of rice genotypes to elevated temperature. Front Plant Sci 6:1044–1055

Sanchez-Reinoso AD, Garces-Varon G, Restrepo-Diaz H (2014) Biochemical and physiological characterization of three rice cultivars under different daytime temperature conditions. Chil J Agric Res 74(4):373–379

Satake T, Yoshida S (1978) High temperature induced sterility in indica rice at flowering. Jpn J Crop Sci 47:6–17

Shah F, Huang J, Kul K, Nie L, Shah T, Chen C (2011) Impact of high-temperature stress on rice plant and its traits related to tolerance. J Agric Sci 149:545–556. https://doi.org/10.1017/S0021859611000360

Shinozaki K, Yamaguchi-Shinozaki K (2007) Gene networks involved in drought stress response and tolerance. J Exp Bot 58:221–227

Shu QY, Wu DX, Xia Y (1997) The most widely cultivated rice variety 'Zhefu 802' in China and its geneology. Mut Breed Newsl 43:3–5

Singla SL, Pareek A, Grover A (1997) High temperature in rice. In: Prasad MNV (ed) Plant ecophysiology. Wiley, New York, pp 101–127

Song JY, Kim DS, Lee MC, Lee KJ, Kim JB, Kim SH, Ha BK, Yun SJ, Kang SY (2012) Physiological characterization of gamma-ray induced salt tolerant rice mutants. Aust J Crop Sci 6:421–429

Takahashi S, Miyazaki M, Matsuo K, Suriyasak C, Tarnada A, Matsuyama K, Iwaya-Inoque M, Ishibashi Y (2016) Differential responses to high temperature during maturation in heat-stress-tolerant cultivars of Japonica rice. Plant Prod Sci 19:2016–2012

Tenorio FA, Ye C, Redona E, Sierra S, Laza M, Argayoso MA (2013) Screening rice genetic resources for heat tolerance. SABRAO J Breed Genet 45:371–381

Tian X, Luo H, Zhou H, Wu C (2009) Research on heat stress of rice in China: progress and prospect. Chin Agric Sci Bull 25:166–168

Tonini A, Cabrera E (2011) Globalizing rice research for a changing world (Technical Bulletin No. 15). International Rice Research Institute, Los Banos

Wahid A, Gelani S, Ashraf M, Foolad MR (2007) Heat tolerance in plants: an overview. Environ Exp Bot 61:199–233

Wassmann R, Jagadish SVK, Sumfleth K, Pathak H, Howell G, Ismail A, Serraj R, Redoña E, Singh RK, Heuer S (2009a) Regional vulnerability of climate change impacts on Asian rice production and scope for adaptation. Adv Agron 102:93–105

Wassmann R, Jagadish SVK, Heuer S, Ismail A, Redona E, Serraj R (2009b) Climate change affecting rice production: the physiological and agronomic basis for possible adaptation strategies. In: Sparks DL (ed) Advances in agronomy, vol 101. Academic, Burlington, pp 59–122

Weerakoon WMW, Maruyama A, Ohba K (2008) Impact of humidity on temperature-induced grain sterility in rice (*Oryza sativa* L). J Agron Crop Sci 194:135–140

Wei-hun Z, Da-wie X, Guo-ping Z (2012) Identification and physiological characterization of thermo-tolerant rice genotypes. J Zhejiang Univ, Agri & Life Sci 38(1):1–9

Wu Y, Chang SJ, Lui HS (2016) Effects of high temperature on grain yield and quality of a subtropical japonica rice- Pon Lai rice. Plant Prod Sci 19(1):145–153

Xia MY, Qi HX (2004) Effects of high temperature on the seed setting percent of hybrid rice bred with four male sterile lines. Hubei Agric Sci 2:21–22

Xue DW, Jiang H, Hu J, Zhang XQ, Guo LB, Zeng DL, Dong GJ, Sun GC, Qian Q (2012) Characterization of physiological response and identification of associated genes under heat stress in rice seedlings. Plant Physiol Biochem 61:46–53

Yang HC, Huang ZO, Jiang ZY, Wang XW (2004) High temperature damage and its protective technologies of early and middle season rice in Anhui province. Anhui Agric Sci 32:3–4

Ye C, Tenerio FA, Argayoso MA, Laza MA, Koh HJ, Redeno ED, Jagadish SVK, Gregorio GB (2015) Identifying and confirming quantitative trait loci associated with heat tolerance at flowering stage in different rice populations. BMC Genet 16:41. https://bmcgenet.biomedcentral.com/articles/10.1186/s12863-015-0199-7

Yoshida S (1981) Fundamentals of rice crops science. International Rice Research Institute, Los Banos, The Philippines

Yoshida S, Satake T, Mackill DS (1981) High temperature stress in rice (review). IRRI Research Paper Series 67

Zhang GL, Chen LY, Zhang ST, Zheng H, Liu GH (2009) Effects of high temperature stress on microscopic and ultrastructural characteristics of mesophyll cells in flag leaves of rice. Ric Sci 16:65–71

Zhao L, Lei J, Huang Y, Zhu S, Chen H, Huang R, Peng Z, Tu Q, Shen X, Yan S (2016) Mapping quantitative trait loci for heat tolerance at anthesis in rice using chromosomal segment substitution lines. Breed Sci 66(3):358–366

Zhong-Hua W, Xin-Chen, Yu-Lin J (2014) Development and characterization of rice mutants for functional genomic studies and breeding. Rice Sci:307–332. http://www.wageningenacademic.com/doi/10.3920/978-90-8686-787-5_16

Zivy M, Thiellement H, Vienne D, Hofmann JP (1983) Study on nuclear and cytoplasmic genomic expression in wheat by two-dimensional gel electrophoresis. Theor Appl Genet 66:1–7

www.ingramcontent.com/pod-product-compliance
Lightning Source LLC
LaVergne TN
LVHW082240200926
848484LV00008B/2381